OFFICIAL SQA PAST PAPERS

WITH ANSWERS

INTERMEDIATE 1 | UNITS 1, 2 & APPLICATIONS

MATHEMATICS
2006-2010

2006 EXAM – page 3
Paper 1 (Non-calculator) – Paper 2

2007 EXAM – page 31
Paper 1 (Non-calculator) – Paper 2

2008 EXAM – page 61
Paper 1 (Non-calculator) – Paper 2

2009 EXAM – page 91
Paper 1 (Non-calculator) – Paper 2

2010 EXAM – page 121
Paper 1 (Non-calculator) – Paper 2

SQA

BrightRED
PUBLISHING

© Scottish Qualifications Authority

First exam published in 2006.
Published by Bright Red Publishing Ltd, 6 Stafford Street, Edinburgh EH3 7AU
tel: 0131 220 5804 fax: 0131 220 6710 info@brightredpublishing.co.uk www.brightredpublishing.co.uk

ISBN 978-1-84948-113-7

A CIP Catalogue record for this book is available from the British Library.

Bright Red Publishing is grateful to the copyright holders, as credited on the final page of the book, for permission to use their material.
Every effort has been made to trace the copyright holders and to obtain their permission for the use of copyright material.
Bright Red Publishing will be happy to receive information allowing us to rectify any error or omission in future editions.

INTERMEDIATE 1

2006

[BLANK PAGE]

FOR OFFICIAL USE

Total mark

X101/102

NATIONAL
QUALIFICATIONS
2006

FRIDAY, 19 MAY
1.00 PM – 1.35 PM

MATHEMATICS
INTERMEDIATE 1
Units 1, 2 and
Applications of Mathematics
Paper 1
(Non-calculator)

Fill in these boxes and read what is printed below.

Full name of centre

Town

Forename(s)

Surname

Date of birth

Day Month Year

Scottish candidate number

Number of seat

1 You may **NOT** use a calculator.

2 Write your working and answers in the spaces provided. Additional space is provided at the end of this question-answer book for use if required. If you use this space, write clearly the number of the question involved.

3 Full credit will be given only where the solution contains appropriate working.

4 Before leaving the examination room you must give this book to the invigilator. If you do not you may lose all the marks for this paper.

SCOTTISH
QUALIFICATIONS
AUTHORITY

©

FORMULAE LIST

Circumference of a circle: $C = \pi d$
Area of a circle: $A = \pi r^2$
Curved surface area of a cylinder: $A = 2\pi rh$

Theorem of Pythagoras:

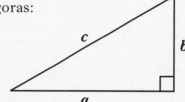

$$a^2 + b^2 = c^2$$

Marks

ALL questions should be attempted.

1. Find $5{\cdot}42 - 1{\cdot}8$.

1

2. A tree surgeon uses this rule to work out his charge in pounds for uprooting and removing trees.

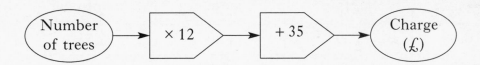

How much would he charge to uproot and remove 11 trees?

2

[Turn over

Marks

3. Paula runs a 1500 metre race at an average speed of 6 metres per second.

   How long does she take to run the race?

   Give her time in minutes and seconds.

3

4. The table below shows insurance premiums for holidays abroad.

	INSURANCE PREMIUM per adult		
	Europe	Worldwide	Winter Sports
Up to 8 days	£15	£30	£40
9–17 days	£20	£40	£55
18–26 days	£30	£60	£80

Child premium (0–15 years) is 70% of the adult premium.

Mr and Mrs Fleming and their 5 year old son go to the USA for a three week holiday in July.

Find the **total** insurance premium for the family.

3

Marks

5. The hire purchase price of this camcorder is £499.

How much will each payment be?

3

[Turn over

Marks

6. The table below shows the **monthly repayments** to be made when money is borrowed from a finance company.

Repayments can be made with or without loan protection.

Loan amount	24 months		36 months		48 months	
	With loan protection	Without loan protection	With loan protection	Without loan protection	With loan protection	Without loan protection
£15 000	£740	£668	£514	£458	£414	£355
£10 000	£493	£445	£343	£304	£276	£237
£ 8000	£394	£356	£274	£244	£220	£189
£ 5000	£246	£222	£171	£152	£138	£118

(a) Johann borrows £10 000 over 4 years **with loan protection**.

How much is his monthly repayment?

1

(b) After 40 months Johann loses his job.

The finance company makes the rest of the repayments for him.

How much does the finance company pay?

2

Marks

7. Ali is a baker. His payslip for the week ending 5th May is shown below. There are some missing entries.

Name: Ali Hackram		Week ending: 5/5/06	
Basic Pay	**Overtime**	**Bonus**	**Gross Pay**
£262·80	£32·85	£20·00	
Income Tax	**National Insurance**	**Pension**	**Total Deductions**
£43·07		£18·94	
			Net Pay
			£225·43

Calculate Ali's:

(*a*) gross pay;

1

(*b*) total deductions;

1

(*c*) national insurance.

2

[Turn over

Marks

8. A television programme has a phone-in to raise money for charity.

    The calls cost 70 pence per minute.

    The charity receives $\frac{3}{5}$ of the cost of each call.

    How much money will the charity receive from a call which lasts $2\frac{1}{2}$ minutes?

3

9. A toilet roll holder is in the shape of a cylinder with diameter 4 centimetres and length 10 centimetres.

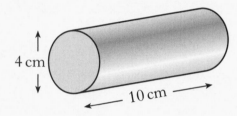

    Calculate the curved surface area of the holder.

    Use π = 3·14.

3

Marks

10. This is a number cell.

	1st	2nd	3rd	4th
	3	−2	1	−1

- 1st number + 2nd number = 3rd number $\quad$ 3 + (−2) = 1
- 2nd number + 3rd number = 4th number $\qquad$ (−2) + 1 = −1

(a) Complete this number cell.

4	−6		

1

(b) Complete this number cell.

		−1	4

2

(c) Complete this number cell.

1			−7

2

YOU MAY USE THE BLANK NUMBER CELLS BELOW FOR WORKING IF YOU WISH.

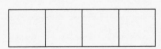

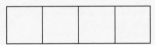

[END OF QUESTION PAPER]

ADDITIONAL SPACE FOR ANSWERS

FOR OFFICIAL USE

Total
mark

X101/104

NATIONAL
QUALIFICATIONS
2006

FRIDAY, 19 MAY
1.55 PM – 2.50 PM

MATHEMATICS
INTERMEDIATE 1
Units 1, 2 and
Applications of Mathematics
Paper 2

Fill in these boxes and read what is printed below.

Full name of centre

Town

Forename(s)

Surname

Date of birth
Day Month Year   Scottish candidate number   Number of seat

1 **You may use a calculator.**

2 Write your working and answers in the spaces provided. Additional space is provided at the end of this question-answer book for use if required. If you use this space, write clearly the number of the question involved.

3 Full credit will be given only where the solution contains appropriate working.

4 Before leaving the examination room you must give this book to the invigilator. If you do not you may lose all the marks for this paper.

SCOTTISH
QUALIFICATIONS
AUTHORITY

FORMULAE LIST

Circumference of a circle: $C = \pi d$
Area of a circle: $A = \pi r^2$
Curved surface area of a cylinder: $A = 2\pi rh$

Theorem of Pythagoras:

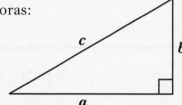

$a^2 + b^2 = c^2$

Marks

ALL questions should be attempted.

1. During a holiday in Mexico, Lee changed £650 into pesos.

   The exchange rate was £1 = 19·13 pesos.

   How many pesos did Lee receive for £650?

   Round off your answer to the nearest ten pesos.

2

[Turn over

Marks

2. The sail of a yacht is sketched below.

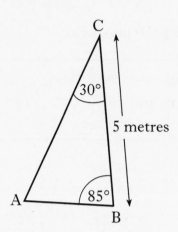

(a) Make a scale drawing of the sail.
   Use a scale of 1 cm to 50 cm.

2

(b) Use your scale drawing to find the actual length of AB, the lower edge of the sail.

   Give your answer in metres.

2

Marks

3. The ages of nine workers in an office are shown below.

23   34   51   19   31   43   38   40   47

Complete the boxplot, drawn below, to show the ages of the workers.

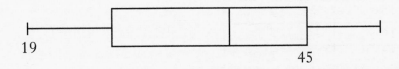

19

45

4

4. The number of bricks needed to build a wall is proportional to the area of the wall.

A wall with an area of 4 square metres needs 260 bricks.

How many bricks are needed for a wall with an area of 7 square metres?

2

[Turn over

Marks

5. A group of 40 students sit a test.

The marks scored by the students in the test are shown in the frequency table below.

Mark	Frequency
14	6
15	10
16	7
17	7
18	5
19	3
20	2

(a) Write down the modal mark.

1

(b) Find the probability of choosing a student from this group with a mark of 19.

1

(c) Complete the table below and calculate the mean mark for the group.

Mark	Frequency	Mark × Frequency
14	6	84
15	10	150
16	7	112
17	7	119
18	5	
19	3	
20	2	
	Total = 40	Total =

3

Marks

6. A water tank is 50 centimetres wide, 1·2 metres long and 40 centimetres high. Calculate its volume.

Give your answer in litres.

(1 litre = 1000 cubic centimetres.)

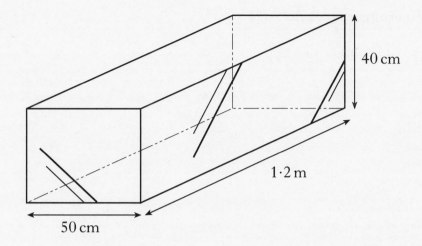

3

[Turn over

Marks

7. Every morning for one week, Wellburgh Council carried out a traffic survey at a busy junction.

The number of cars waiting to turn right at the junction was counted every five minutes between 8 am and 9 am.

On Monday morning the results were:

    10   14   17   12   14   11   13   7   8   7   6   2.

Calculate:

(*a*) the median;

2

(*b*) the range.

2

On Saturday morning, the median was 6 and the range was 8.

(*c*) Make **two** comments comparing the number of cars waiting to turn right at the junction on Monday morning and Saturday morning.

2

Marks

8. Stephen is playing snooker.

   The diagram below shows the positions of three balls on the table.

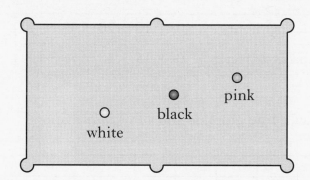

Stephen plays the white ball, W.

It bounces off the side of the table at X and hits the pink ball, P.

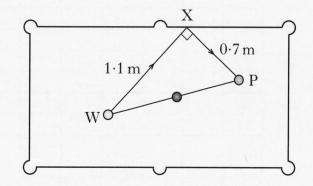

- Distance WX is 1·1 metres
- Distance XP is 0·7 metres
- Angle WXP is 90°

Calculate distance WP.

Do not use a scale drawing.

3

[Turn over

Marks

9. The table below shows the stopping distances of a car, when the brakes are applied, at different speeds.

Speed (miles per hour)	0	10	20	30	40
Stopping distance (feet)	0	15	40	75	120

On the grid below, draw a **line** graph to show this information.

4

DO NOT
WRITE IN
THIS
MARGIN

Marks

10. Ralph invests £2600 in a building society account.

The rate of interest is 4·5% per annum.

Calculate the interest he should receive after 8 months.

3

[Turn over

Marks

11. A supplier gives shopkeepers a discount on large orders of bottled water.
The flowchart below is used to work out the discount in pounds.

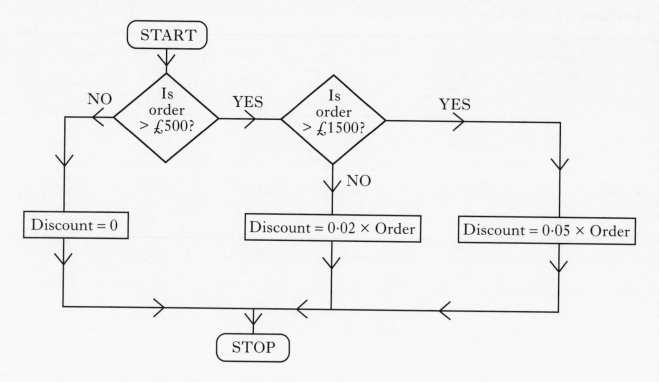

(a) A shopkeeper orders £800 worth of bottled water from this supplier **every week**.

How much does she pay each week for this order?

2

(b) The shopkeeper decides to order £1600 worth of bottled water **once every fortnight** instead of an £800 order every week.

How much **less** will she pay every fortnight?

3

Marks

12. Andrew designs a website to advertise his hotel.

    In the first month he has 250 visitors to his site.

    The following month he has 300 visitors.

    Calculate the percentage increase in the number of visitors.

4

[Turn over for Question 13 on *Page fourteen*

Marks

13. The diagram below shows the wall at the start of a tunnel.

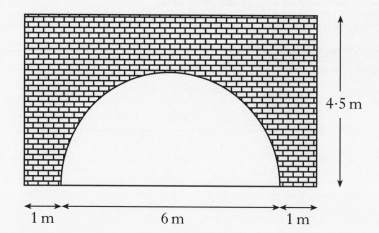

4·5 m

1 m   6 m   1 m

The wall is in the shape of a rectangle with a semi-circular space for the tunnel.

Calculate the area of the wall in square metres.

Give your answer correct to one decimal place.

5

[END OF QUESTION PAPER]

ADDITIONAL SPACE FOR ANSWERS

DO NOT
WRITE I
THIS
MARGIN

ADDITIONAL SPACE FOR ANSWERS

[BLANK PAGE]

FOR OFFICIAL USE

Total mark

X101/102

NATIONAL
QUALIFICATIONS
2007

TUESDAY, 15 MAY
1.00 PM – 1.35 PM

MATHEMATICS
INTERMEDIATE 1
Units 1, 2 and
Applications of Mathematics
Paper 1
(Non-calculator)

Fill in these boxes and read what is printed below.

Full name of centre

Town

Forename(s)

Surname

Date of birth
Day Month Year   Scottish candidate number   Number of seat

1 You may **NOT** use a calculator.

2 Write your working and answers in the spaces provided. Additional space is provided at the end of this question-answer book for use if required. If you use this space, write clearly the number of the question involved.

3 Full credit will be given only where the solution contains appropriate working.

4 Before leaving the examination room you must give this book to the invigilator. If you do not you may lose all the marks for this paper.

SCOTTISH
QUALIFICATIONS
AUTHORITY

FORMULAE LIST

Circumference of a circle: $C = \pi d$

Area of a circle: $A = \pi r^2$

Curved surface area of a cylinder: $A = 2\pi rh$

Theorem of Pythagoras:

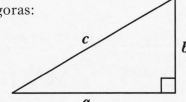

$a^2 + b^2 = c^2$

ALL questions should be attempted.

Marks

1. (*a*) Find $8 \cdot 52 + 10 \cdot 7$.

1

(*b*) Find $3 \cdot 76 \div 8$.

1

(*c*) Change $0 \cdot 057$ into a fraction.

1

(*d*) Find 90% of £320.

2

2. Shona wants to insure her jewellery for £8000.

The insurance company charges an annual premium of £7·65 for each £1000 insured.

Work out Shona's annual premium.

2

Marks

3. The network diagram shows the time it takes to make a meal of Spaghetti Bolognese.

All times are in minutes.

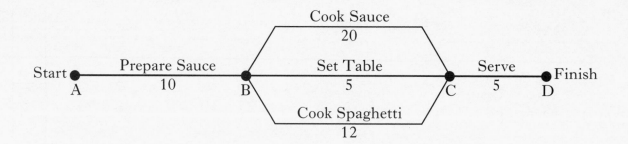

(*a*) How long does it take to cook the spaghetti?

1

(*b*) How long does it take altogether to make the meal from start to finish?

1

Marks

4. The number of minutes that patients had to sit in the waiting room before seeing their doctor was recorded one day.

The results are shown in the frequency table below.

Number of minutes	Frequency	Number of minutes × Frequency
5	4	20
6	7	42
7	8	56
8	13	104
9	12	
10	6	
	Total = 50	Total =

Complete the table above **and** find the mean number of minutes.

3

[Turn over

Marks

5. The diagram below shows the net of a triangular prism.

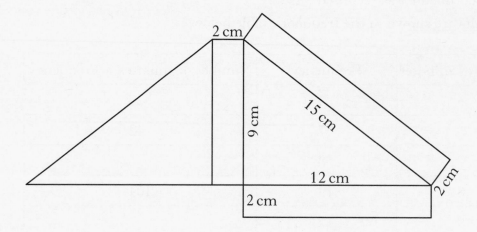

Find the total **surface area** of the triangular prism.

3

Marks

6. Shown below is a container in the shape of a cuboid.

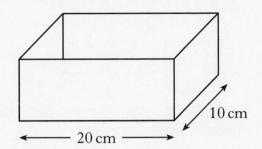

10 cm

20 cm

When full, the container holds 1600 cubic centimetres of water.

Work out the height of the container.

3

7. Work out the answers to the following.

(*a*) $2 \times (-2) \times 2$

1

(*b*) $11 - (-6)$

1

[**Turn over**

Marks

8. Naveed has six electrical appliances in his student lodgings.

   The power, in watts, used by each appliance is shown below.

Lamp 100 watts

Computer 200 watts

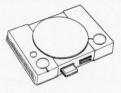

Games Machine
400 watts

Microwave 700 watts

Heater 1000 watts

Kettle 2300 watts

Naveed uses a 4-way extension lead for the appliances.

The instructions state that the maximum power used through the extension lead should not be more than 3000 watts.

One combination of **four** appliances that Naveed can safely use through the extension lead is shown in the table below.

Lamp 100 watts	Computer 200 watts	Games Machine 400 watts	Microwave 700 watts	Heater 1000 watts	Kettle 2300 watts	Total Watts
✓	✓	✓		✓		1700

Complete the table to show **all** the possible combinations of **four** appliances that Naveed can safely use through the extension lead.

3

Marks

9. The number of goals scored by each of twelve football teams in a season is shown below.

   38   33   35   57   60   53   50   52   55   73   80   62

   (a) Find the upper quartile.

   2

   (b) Calculate the interquartile range.

   2

10. Black and white counters are placed in two bags as shown below.

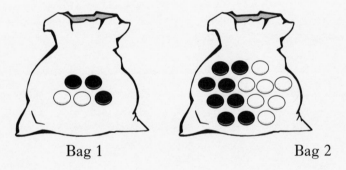

Bag 1   Bag 2

One counter is selected at random from **each** bag.

Which bag gives a greater probability of selecting a black counter?

Explain your answer.

3

[END OF QUESTION PAPER]

ADDITIONAL SPACE FOR ANSWERS

ADDITIONAL SPACE FOR ANSWERS

ADDITIONAL SPACE FOR ANSWERS

ADDITIONAL SPACE FOR ANSWERS

FOR OFFICIAL USE

Total
mark

X101/104

NATIONAL
QUALIFICATIONS
2007

TUESDAY, 15 MAY
1.55 PM – 2.50 PM

MATHEMATICS
INTERMEDIATE 1
Units 1, 2 and
Applications of Mathematics
Paper 2

Fill in these boxes and read what is printed below.

Full name of centre

Town

Forename(s)

Surname

Date of birth

Day   Month   Year   Scottish candidate number   Number of seat

1   **You may use a calculator.**

2   Write your working and answers in the spaces provided. Additional space is provided at the end of this question-answer book for use if required. If you use this space, write clearly the number of the question involved.

3   Full credit will be given only where the solution contains appropriate working.

4   Before leaving the examination room you must give this book to the invigilator. If you do not you may lose all the marks for this paper.

SCOTTISH
QUALIFICATIONS
AUTHORITY

©

FORMULAE LIST

Circumference of a circle: $C = \pi d$

Area of a circle: $A = \pi r^2$

Curved surface area of a cylinder: $A = 2\pi rh$

Theorem of Pythagoras:

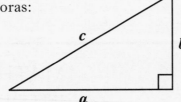

$$a^2 + b^2 = c^2$$

ALL questions should be attempted.

1. The bar graph shows the number of hotels in Southbay awarded grades A to E by the local tourist board.

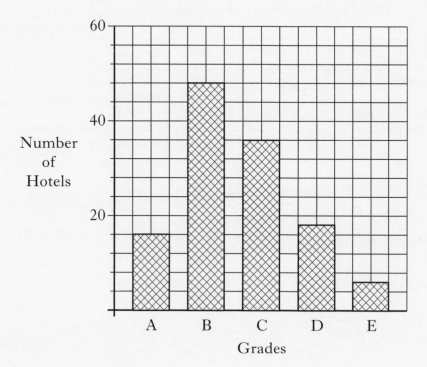

(a) How many hotels were awarded an A grade?

1

(b) Write down the modal grade.

1

[Turn over

DO NOT
WRITE I
THIS
MARGI

Marks

2. The distance travelled (in miles) by a lorry driver each day is recorded in the spreadsheet below.

	A	B	C	D	E	F	G
1		Monday	Tuesday	Wednesday	Thursday	Friday	Total Distance
2	Week 1	110	80	90	80	60	
3	Week 2	100	90	100	70	70	
4	Week 3	90	100	90	100	80	
5	Week 4	120	90	80	90	70	
6							

(a) What formula would be used to enter the total distance travelled in week 1 in cell G2?

1

(b) The result of the formula =AVERAGE(B2..B5) is to be entered in cell B6.

What would appear in cell B6?

1

3. An aeroplane took off from Edinburgh at 0753 and landed in Shetland at 0908. The distance flown by the aeroplane was 295 miles.

Calculate the average speed of the aeroplane in miles per hour.

3

Marks

4. Wayne is a mechanic.

He earns £329·70 for a basic 35 hour week.

(*a*) Calculate his hourly rate.

1

(*b*) One week Wayne also works 3 hours overtime.

His overtime rate is time and a half.

How much does he earn **altogether** for that week?

3

[Turn over

Marks

5. A teacher records the number of absences and end of term test mark for each of her students.

The scattergraph shows the results.

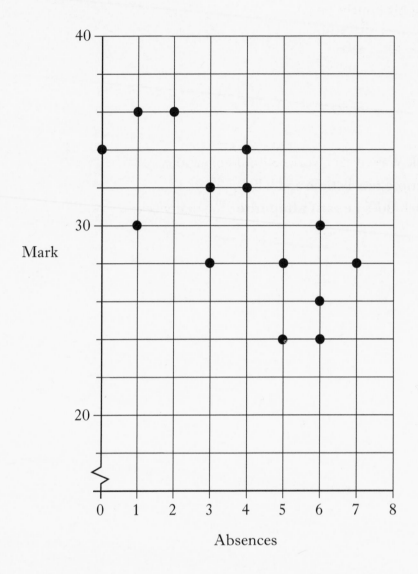

(*a*) Draw a line of best fit through the points on the graph.   **1**

(*b*) Use your line of best fit to estimate the mark of a student who had 8 absences.

1

Marks

6. The table below shows the **monthly payments** to be made when money is borrowed from a loan company.

   Payments can be made with or without payment protection.

	WITHOUT PAYMENT PROTECTION		
Loan Amount	12 months	36 months	60 months
£ 2000	£184·47	£70·56	£48·25
£ 5000	£442·61	£160·35	£104·31
£10 000	£882·94	£318·76	£206·68
£15 000	£1324·41	£478·14	£310·02

	WITH PAYMENT PROTECTION		
Loan Amount	12 months	36 months	60 months
£ 2000	£200·38	£87·08	£66·68
£ 5000	£480·78	£197·88	£144·15
£10 000	£959·08	£393·38	£265·62
£15 000	£1438·63	£590·06	£428·42

Gail borrows £5000 over 3 years **with payment protection**.

(a) State her monthly payment.

1

(b) Calculate her total payments.

1

(c) Calculate how much this loan cost Gail.

1

[Turn over

Marks

7. The weights of two groups of ten people are to be compared.
    Listed below are the weights (in kilograms) of the ten people in group A.

    64   71   73   66   69   78   77   75   76   71

    (a) Find the median.

    2

    (b) Find the range.

    2

    (c) For the ten people in group B the median is 76 and the range is 20.
         Make **two** comments comparing the weights of the people in group A
         and group B.

    2

Marks

8. Sam invests £7600 in a bank account.

- The rate of interest is 4·8% per annum.
- The bank deducts 20% tax from the interest.

Calculate the interest Sam receives for one year after tax has been deducted.

3

[Turn over

Marks

9. Phil is making a wooden bed frame.

    The frame is rectangular and measures 195 centimetres by 95 centimetres.

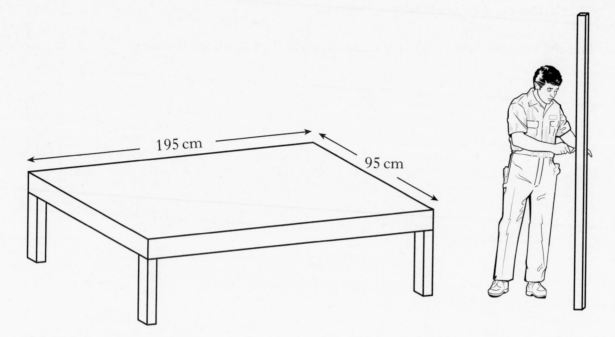

    To make the frame rigid, Phil is going to add a piece of wood along one of its diagonals.

    He has a piece of wood 2·2 metres long.

    Is this piece of wood long enough to fit along the diagonal?

    Give a reason for your answer.

    Do not use a scale drawing.

4

Marks

10. Curtis flew from New York to London where he changed 1400 dollars into pounds.

He spent £650 in London and then changed the rest into euros before travelling to Paris.

How many euros did Curtis receive?

Exchange Rates	
$	£1 = 1·75 dollars
€	£1 = 1·38 euros

3

[Turn over

Marks

11. Jill is taking part in an orienteering competition.

She starts at checkpoint A as shown below.

She then runs due east for 900 metres to checkpoint B.

(a) Show the position of checkpoint B in a scale drawing.

Use a scale of 1 cm to 100 m.

N

A

1

(b) Checkpoint C lies on a bearing of:

- 055° from checkpoint A
- 320° from checkpoint B.

Complete the scale drawing to show the position of checkpoint C.   3

Marks

12. Pamela paid £40 for a concert ticket.

She was unable to go to the concert, so she sold her ticket on the Internet for £26.

Express her loss as a percentage of what she paid for the ticket.

4

[Turn over

Marks

13.   The diagram below shows a birthday card.

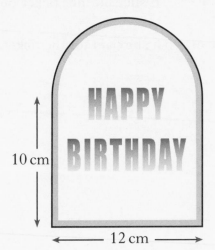

The card consists of a rectangle and a semi-circle.

There is gold ribbon all round the border of the card.

Calculate the total length of gold ribbon needed for this card.

Give your answer to the **nearest centimetre**.

5

Marks

14. The tariffs shown below are available when buying a mobile phone.

Pay As You Go	**Monthly Contract**
Calls: 14p per minute	**Rental:** £18 per month **Calls:** 6p per minute

(*a*) Find the cost of using 200 minutes of calls each month on the:

 (i) Pay As You Go tariff;

 (ii) Monthly Contract tariff.

2

(*b*) Nick and Amy have mobile phones.

Nick is on Pay As You Go and Amy has a Monthly Contract.

In April:

• the cost to each was exactly the same
• Nick used the same number of minutes as Amy.

How many minutes was this?

3

[END OF QUESTION PAPER]

ADDITIONAL SPACE FOR ANSWERS

[BLANK PAGE]

FOR OFFICIAL USE

X101/102

Total mark

NATIONAL
QUALIFICATIONS
2008

TUESDAY, 20 MAY
1.00 PM – 1.35 PM

MATHEMATICS
INTERMEDIATE 1
Units 1, 2 and
Applications of Mathematics
Paper 1
(Non-calculator)

Fill in these boxes and read what is printed below.

Full name of centre

Town

Forename(s)

Surname

Date of birth

Day   Month   Year   Scottish candidate number

Number of seat

1 **You may NOT use a calculator.**

2 Write your working and answers in the spaces provided. Additional space is provided at the end of this question-answer book for use if required. If you use this space, write clearly the number of the question involved.

3 Full credit will be given only where the solution contains appropriate working.

4 Before leaving the examination room you must give this book to the invigilator. If you do not you may lose all the marks for this paper.

Use blue or black ink. Pencil may be used for graphs and diagrams only.

FORMULAE LIST

Circumference of a circle: $C = \pi d$
Area of a circle: $A = \pi r^2$
Curved surface area of a cylinder: $A = 2\pi r h$

Theorem of Pythagoras:

$$a^2 + b^2 = c^2$$

Marks

ALL questions should be attempted.

1. (*a*) Find $2 \cdot 658 - 0 \cdot 29$.

1

(*b*) Find 14×3000.

1

(*c*) Find $5 \cdot 45 \div 5$.

1

2. Sandra works night shift. One night she started work at 2235 and finished at 0715 the next morning.

How long did Sandra's shift last?

1

[Turn over

Marks

3. The number of salmon caught in a river over a four year period is recorded in the spreadsheet below.

	A	B	C	D	E	F
1		2004	2005	2006	2007	
2	MAY	15	3	0	0	
3	JUNE	139	109	171	234	
4	JULY	267	225	216	276	
5	AUGUST	159	103	72	48	
6	SEPTEMBER	41	13	21	1	
7						

(a) The result of the formula = SUM(E2..E6) is to be entered in cell E7. What would appear in cell E7?

1

(b) What formula would be used to enter the average number of fish caught in June over the four year period in cell F3?

1

4. A plumber charges £20 for being called out to a job, plus £12 **for each 15 minutes** he takes to do the job.

How much does he charge for a job which takes 2 hours?

2

Marks

5. A building company employs 70 staff.

The number of staff absences during the last year is shown in the frequency table below.

Number of Absences (Days)	Frequency
0	7
1	21
2	18
3	11
4	8
5	5
Total	70

(*a*) Find the probability of choosing a member of staff who had no absences.

1

(*b*) Complete the table below **and** calculate the mean number of absences.

Number of Absences (Days)	Frequency	Number of Absences × Frequency
0	7	0
1	21	21
2	18	36
3	11	
4	8	
5	5	
Total	70	

3

6. Frances is on holiday. She wants to book some of the excursions shown in the advert below.

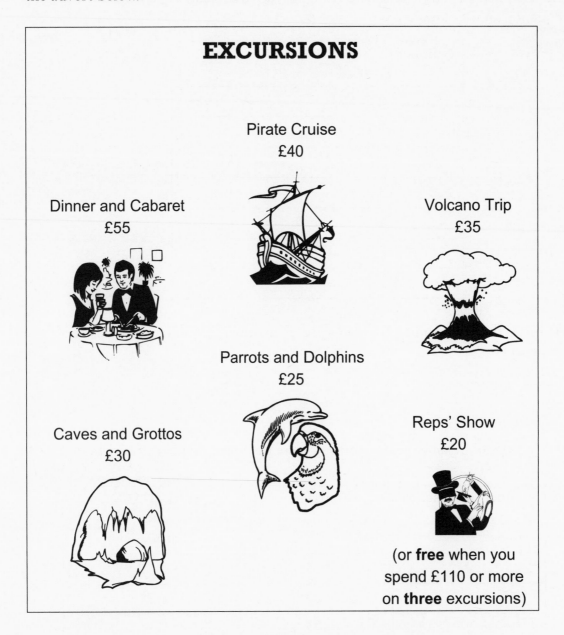

EXCURSIONS

Pirate Cruise
£40

Dinner and Cabaret
£55

Volcano Trip
£35

Parrots and Dolphins
£25

Caves and Grottos
£30

Reps' Show
£20

(or **free** when you
spend £110 or more
on **three** excursions)

- Frances wants to book **four** different excursions.
- She can afford to spend a **maximum of £120**.
- She gets a **free** ticket for the Reps' Show when she spends £110 or more on **three** excursions.

Marks

6. (continued)

Two combinations of **four** excursions that Frances can afford are shown in the table below.

Dinner and Cabaret £55	55							
Pirate Cruise £40		40						
Volcano Trip £35		35						
Caves and Grottos £30	30							
Parrots and Dolphins £25	25	25						
Reps' Show £20 or Free	Free	20						
Total Price	£110	£120						

Complete the table to show **all** possible combinations that Frances can afford.

3

[Turn over

Marks

7. A child health survey monitors the ages at which young children can build a tower of four wooden blocks.

The ages (in months) of a group of children are shown below.

23   16   14   20   18   17   16   20   17   19   13   25   24

Complete the boxplot, drawn below, to show this information.

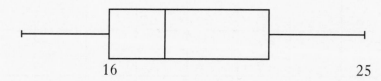

16   25

4

Marks

8. The scale drawing shows the route taken by a ferry from Hayton to Eastport.

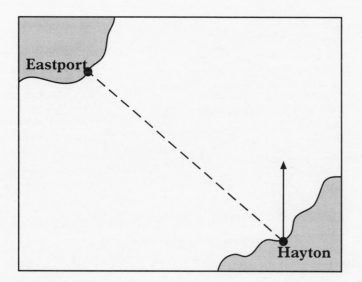

Scale: 1 cm to 250 m

Use the scale drawing to find the distance and bearing of Eastport from Hayton.

4

[Turn over for Questions 9 and 10 on *Page ten*

Marks

9. Evaluate $x^2 - y$ when $x = -8$ and $y = 73$.

3

10. Jamie invests £1440 in a savings account.

    The rate of interest is 5% per annum.

    Calculate the interest he should receive after 3 months.

4

[END OF QUESTION PAPER]

ADDITIONAL SPACE FOR ANSWERS

Page eleven

ADDITIONAL SPACE FOR ANSWERS

[BLANK PAGE]

FOR OFFICIAL USE

Total mark

X101/104

NATIONAL
QUALIFICATIONS
2008

TUESDAY, 20 MAY
1.55 PM – 2.50 PM

MATHEMATICS
INTERMEDIATE 1
Units 1, 2 and
Applications of Mathematics
Paper 2

Fill in these boxes and read what is printed below.

Full name of centre

Town

Forename(s)

Surname

Date of birth

Day	Month	Year

Scottish candidate number

Number of seat

1   **You may use a calculator.**

2   Write your working and answers in the spaces provided. Additional space is provided at the end of this question-answer book for use if required. If you use this space, write clearly the number of the question involved.

3   Full credit will be given only where the solution contains appropriate working.

4   Before leaving the examination room you must give this book to the invigilator. If you do not you may lose all the marks for this paper.

Use blue or black ink. Pencil may be used for graphs and diagrams only.

FORMULAE LIST

Circumference of a circle: $C = \pi d$
Area of a circle: $A = \pi r^2$
Curved surface area of a cylinder: $A = 2\pi rh$

Theorem of Pythagoras:

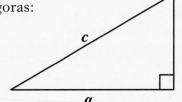

$$a^2 + b^2 = c^2$$

Marks

ALL questions should be attempted.

1. (*a*) On the grid below plot the points A(–2,4), B(–4,–1) and C(1,–3).

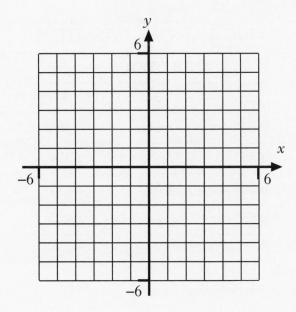

2

(*b*) Plot the point D so that shape ABCD is a square.

1

[Turn over

Marks

2. The table below shows the basic annual premiums charged for car insurance by an insurance company.

The basic premium depends on the area where the driver lives and the group their car belongs to.

AREA	BASIC ANNUAL PREMIUM				
	CAR GROUP				
	1	2	3	4	5
A	£428	£517	£613	£725	£838
B	£497	£555	£659	£779	£898
C	£525	£598	£712	£841	£975
D	£540	£651	£775	£915	£1055

(a) Lynn's car is in group 4 and she lives in area C.

Write down her basic annual premium.

1

Drivers who do not make a claim on their insurance receive a discount on their basic annual premium as shown in the table below.

Number of years without a claim	1	2	3	4 or more
Discount	30%	40%	55%	67%

(b) Lynn has not made a claim for 4 years.

How much will it cost her to insure her car?

2

Marks

3. A group of friends are booking some rooms at the Westcliff Hotel for a short holiday. This flowchart is used to work out the cost of rooms at the hotel.

The group book **6 rooms** for **4 nights**.

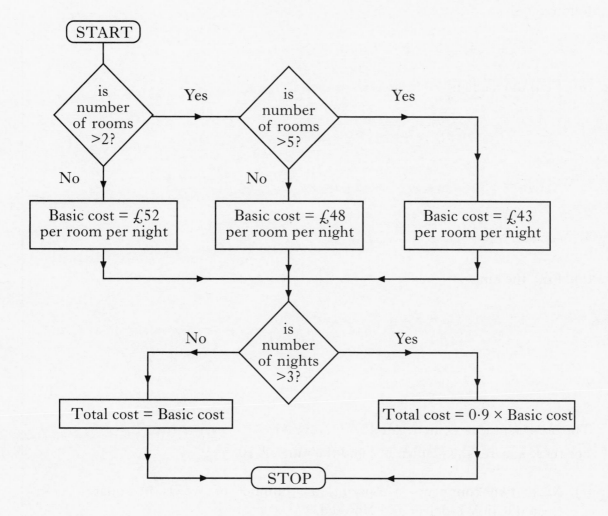

Work out the total cost of **6 rooms** for **4 nights**.

3

Marks

4. A grass lawn is treated with weedkiller.

The lawn is split into twenty squares each of the same area.

Ten of the squares are treated with Weedclear.

Three weeks later the number of weeds in each of these squares is as follows:

$$3, \quad 4, \quad 6, \quad 2, \quad 1, \quad 7, \quad 2, \quad 1, \quad 1, \quad 3.$$

(*a*) Find the median.

2

(*b*) Find the range.

1

The other ten squares are treated with Noweed.

For these squares the median is 2 and the range is 10.

(*c*) Make **two** comments comparing the number of weeds in squares treated with Weedclear and Noweed.

2

Marks

5. Ross drove 190 miles from Preston to Edinburgh in 3 hours 30 minutes.

   During the first part of his journey he drove for 2 hours at an average speed of 68 miles per hour.

   Find the average speed in miles per hour for the rest of his journey.

4

[Turn over

Marks

6. Some biology students were doing a project on "creepy crawlies". The pie chart shows the different types of creepy crawlies that the students collected from a garden.

CREEPY CRAWLIES

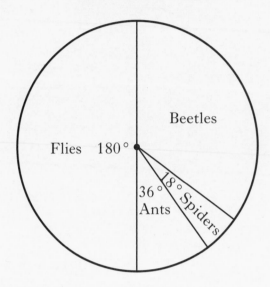

The students collected 220 creepy crawlies altogether.

How many of them were beetles?

3

Marks

7. A farmer is building a wire fence around a field.

   The fence has heavy posts at the corners.

   Each corner post is supported by a stake as shown in the diagram.

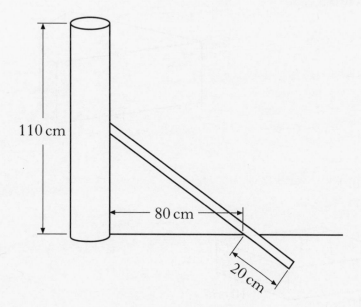

   - The corner post is 110 centimetres high.
   - The stake meets the corner post halfway up.
   - The stake meets the ground 80 centimetres from the foot of the corner post.
   - 20 centimetres of the stake is below ground level.

   Calculate the length of the stake.

   Do not use a scale drawing.

   4

   [Turn over

Marks

8. Shown below are two pieces of cheese.

   The weight of each piece is proportional to its volume.

   Piece A has a volume of 400 cubic centimetres.
   It weighs 480 grams.

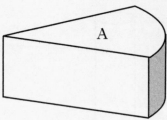

   Piece B is a cuboid.

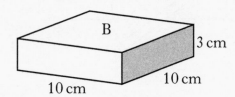

   Find the weight of piece B.

4

Marks

9. The table shows the ticket prices for a theme park in France.

    The prices are given in euros.

Ticket		Adult price	Child price
Bronze	(valid 1 day)	€50	€40
Silver	(valid 2 days)	€90	€75
Gold	(valid 3 days)	€110	€85

Gavin buys silver tickets for two adults and one child.

Find the total cost, in pounds and pence, of buying these tickets if the exchange rate is £1 = 1·39 euros.

3

[Turn over

Marks

10. Danny borrows £1700 from a credit company.

The loan has to be repaid in twelve months.

The loan can be repaid by making

    either • twelve monthly payments

    or • a single payment at the end of the twelve months.

The interest rates charged by the company are shown below.

INTEREST RATES	
Pay Monthly monthly interest rate 1·6%	OR **Single Payment** **at end of 12 months** annual percentage rate (APR) 21%

(*a*) Danny considers making monthly payments.

How much interest would be charged for the first month?

1

(*b*) Danny decides to make the single payment at the end of twelve months.

How much is this payment?

2

Marks

11. Bernie works in an office.

He earns £8·20 per hour for a basic 35 hour week.

One weekend he works overtime as shown in the table below.

	Time and a Half	Double Time
Saturday	9 am – 12 noon	1 pm – 4 pm
Sunday	—	9 am – 1 pm

How much does Bernie earn **altogether** for this week?

4

[Turn over

DO NO⸱
WRITE I
THIS
MARGI⸱

Marks

12. A cylinder has diameter 18 centimetres and height 2·5 centimetres.

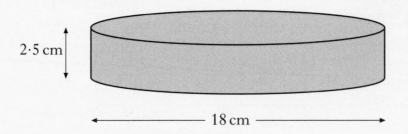

Calculate the **curved** surface area of the cylinder.

3

Marks

13. Sergei has been training to run a marathon.

Since he started training his weight has dropped from 80 kilograms to 74 kilograms.

Express his weight loss as a percentage of his original weight.

4

[Turn over for Question 14 on *Page sixteen*

Marks

14. The diagram below shows part of a garden which is being watered from a sprinkler.

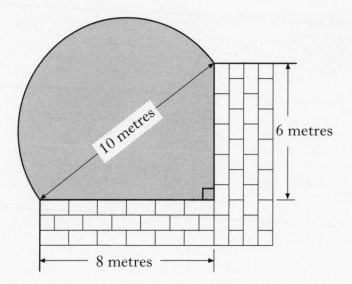

The area being watered is in the shape of a semi-circle and a right angled triangle.

Calculate the area being watered.

4

[END OF QUESTION PAPER]

[BLANK PAGE]

FOR OFFICIAL USE

Total mark

X101/102

NATIONAL
QUALIFICATIONS
2009

THURSDAY, 21 MAY
1.00 PM – 1.35 PM

MATHEMATICS
INTERMEDIATE 1
Units 1, 2 and
Applications of Mathematics
Paper 1
(Non-calculator)

Fill in these boxes and read what is printed below.

Full name of centre

Town

Forename(s)

Surname

Date of birth

Day		Month		Year	

Scottish candidate number

Number of seat

1 **You may NOT use a calculator.**

2 Write your working and answers in the spaces provided. Additional space is provided at the end of this question-answer book for use if required. If you use this space, write clearly the number of the question involved.

3 Full credit will be given only where the solution contains appropriate working.

4 Before leaving the examination room you must give this book to the invigilator. If you do not you may lose all the marks for this paper.

Use blue or black ink. Pencil may be used for graphs and diagrams only.

FORMULAE LIST

Circumference of a circle: $C = \pi d$

Area of a circle: $A = \pi r^2$

Curved surface area of a cylinder: $A = 2\pi r h$

Theorem of Pythagoras:

$$a^2 + b^2 = c^2$$

ALL questions should be attempted.

Marks

1. (*a*) Find $28 \cdot 7 + 4 \cdot 35$.

1

   (*b*) Find $1 \cdot 89 \div 7$.

1

   (*c*) Find $6 \times 4\frac{1}{3}$.

2

2. The cost in pounds of hiring a function room is:

Cost = 250 + (number of people × 4·99)

   Find the cost of hiring the function room for 200 people.

2

[Turn over

Marks

3. Work out the answers to the following.

   (*a*) $9 + (-13)$

   1

   (*b*) $-56 \div (-8)$

   1

4. The cash price of a television is £260.

   It can be bought on hire purchase by paying a deposit of 25% of the cash price and 15 instalments of £14.

   Find the total hire purchase price of the television.

   3

Marks

5. Sally is paid £9·50 per hour. She works 30 hours a week. Complete her payslip below.

Name: Sally Jones		**Week ending: 15/5/09**
Hours worked 30	**Hourly rate** £9·50	**Gross pay**
Tax £37·80	**National Insurance** £21·48	**Total deductions**
		Net pay

3

6. The birth weights (in kilograms) of ten babies born in a hospital in one day are shown below.

    2·4   3·6   2·2   2·9   3·1   3·3   2·7   3·9   3·4   3·8

    (a) Find the lower quartile.

2

    (b) Calculate the interquartile range.

2

Marks

7. A group of 40 visitors to the Edinburgh Festival Fringe were asked how many performances they had attended.

The results are shown in the frequency table below.

Number of performances	Frequency
5	2
6	9
7	11
8	9
9	4
10	4
11	1
	Total = 40

(*a*) Write down the modal number of performances.

1

(*b*) Find the range of the number of performances.

1

(*c*) Complete the table below **and** find the mean number of performances.

Number of performances	Frequency	Number of performances × Frequency
5	2	10
6	9	54
7	11	77
8	9	72
9	4	
10	4	
11	1	
	Total = 40	Total =

3

Marks

8. This network diagram shows the distances between six villages.
All distances are in miles.

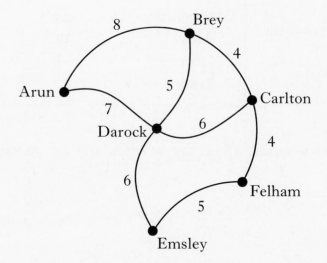

(a) Write down the order of the node at Darock.

1

(b) Complete the distance chart below to show the **shortest** distance between each pair of villages.

Arun					
8	Brey				
	4	Carlton			
7	5	6	Darock		
13	11		6	Emsley	
	8	4		5	Felham

[Turn over for Question 9 on *Page eight*

2

DO NOT
WRITE IN
THIS
MARGIN

Marks

9. In a quiz, three contestants are each asked 20 questions.

The contestants score

- +2 points for a correct answer
- 0 points for no answer
- −1 point for a wrong answer.

SCOREBOARD

Contestant	Points
Amy	10
John	−2
Fred	

(*a*) Fred gives 4 correct answers, 15 wrong answers and no answer to 1 question.

How many points does he score?

1

(*b*) Amy's score for her 20 questions is 10 points.

One way that she can score 10 points is shown in the table below.

Correct answer +2 points	No answer 0 points	Wrong answer −1 point	Total
7	9	4	10
			10
			10
			10
			10
			10

Complete the table to show **all** the possible ways that Amy can score 10 points.

3

[*END OF QUESTION PAPER*]

ADDITIONAL SPACE FOR ANSWERS

ADDITIONAL SPACE FOR ANSWERS

ADDITIONAL SPACE FOR ANSWERS

FOR OFFICIAL USE

X101/104

Total mark

NATIONAL
QUALIFICATIONS
2009

THURSDAY, 21 MAY
1.55 PM – 2.50 PM

MATHEMATICS
INTERMEDIATE 1
Units 1, 2 and
Applications of Mathematics
Paper 2

Fill in these boxes and read what is printed below.

Full name of centre

Town

Forename(s)

Surname

Date of birth

Day   Month   Year   Scottish candidate number

Number of seat

1   **You may use a calculator.**

2   Write your working and answers in the spaces provided. Additional space is provided at the end of this question-answer book for use if required. If you use this space, write clearly the number of the question involved.

3   Full credit will be given only where the solution contains appropriate working.

4   Before leaving the examination room you must give this book to the invigilator. If you do not you may lose all the marks for this paper.

Use blue or black ink. Pencil may be used for graphs and diagrams only.

[BLANK PAGE]

FORMULAE LIST

Circumference of a circle: $C = \pi d$

Area of a circle: $A = \pi r^2$

Curved surface area of a cylinder: $A = 2\pi rh$

Theorem of Pythagoras:

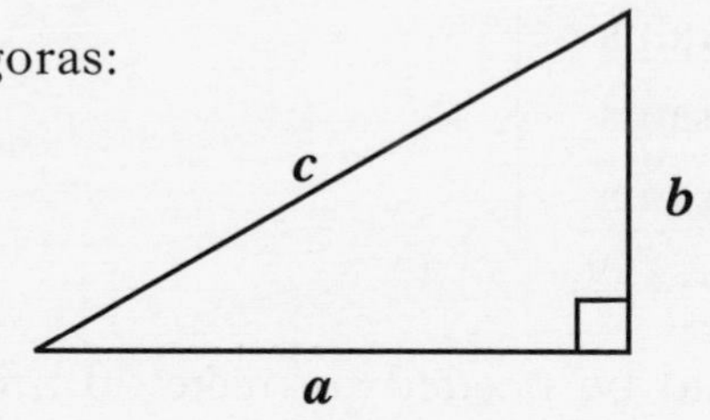

$a^2 + b^2 = c^2$

[Turn over

Marks

ALL questions should be attempted.

1. A recipe lists the ingredients needed to make 8 mincemeat pies.

Ingredients for 8 mincemeat pies	
Plain flour	60 grams
Lard	20 grams
Butter	20 grams
Mincemeat	180 grams

How many grams of plain flour would be needed to make 30 mincemeat pies?

2

2. Stefan uses this spreadsheet to calculate his car's petrol consumption. He records the distance his car travels for each quantity of petrol he buys.

	A	B	C	D
		Petrol (gallons)	Distance (miles)	Petrol Consumption (miles per gallon)
1	Date			
2	1 April	10	330	33
3	6 April	7	210	30
4	12 April	12	384	
5	25 April	8	276	
6	30 April	11	385	
7				

(a) The result of the formula =SUM(B2..B6) is to be entered in cell B7. What would appear in cell B7?

(b) What **formula** would be used to enter the petrol consumption in cell D4?

[Turn over

Marks

3. Jenny takes out a loan for £4500.

   She is charged 14% interest on the amount borrowed.

   She repays the total owed in twelve equal payments.

   How much is each payment?

3

Marks

4. The number of cheese sandwiches sold by a sandwich bar was recorded for 15 days.

28	38	19	33	29
32	37	41	27	50
45	23	44	38	34

(*a*) Display this information in a stem and leaf diagram.

3

(*b*) Find the median number of cheese sandwiches sold.

1

(*c*) Find the probability that more than 40 cheese sandwiches were sold on any day.

1

[Turn over

Marks

5. The time in Glasgow is 5 hours ahead of the time in New York.

    When it is noon in New York, it is 5 pm in Glasgow.

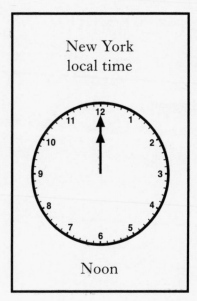
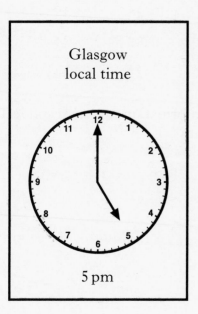

New York
local time

Noon

Glasgow
local time

5 pm

An aeroplane leaves New York at 9.20 pm local time to fly to Glasgow.

It flies 3220 miles at an average speed of 560 miles per hour.

What is the local time in Glasgow when the plane arrives?

4

Marks

6. (*a*) The diagram below shows the net of a triangular prism.

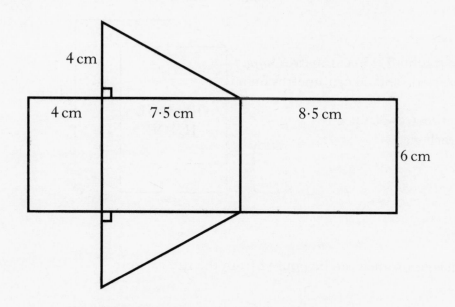

Find the total surface area of the triangular prism.

3

(*b*) This cube has the same surface area as the triangular prism shown above.

Calculate the length of one edge of the cube.

2

[Turn over

Marks

7.   Orange juice is poured from a carton into some glasses.

The carton is a cuboid, 15 centimetres long,
10 centimetres wide and 20 centimetres high.

125 cubic centimetres of juice is
poured into each glass.

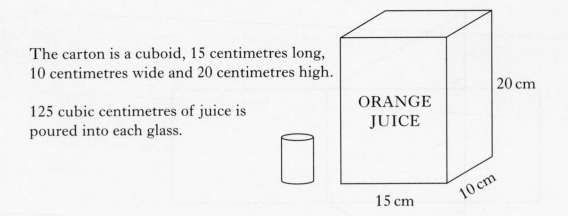

How many glasses of juice can be poured from the full carton?

3

Marks

8. Colin is going on holiday to Spain.

   He wants to exchange a maximum of £1300 into euros.

   The exchange rate is £1 = €1·26.

   His bank only issues euros in multiples of €10.

   (*a*) What is the maximum number of euros Colin will receive from his bank?

   2

   (*b*) How much will Colin actually pay for this number of euros?

   2

   [Turn over

Marks

9. The pie chart shows how the pupils at Newdale Academy travelled to school each day.

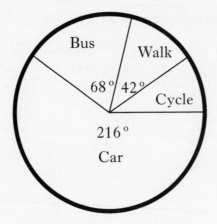

(a) There are 900 pupils at Newdale Academy.

How many pupils cycled to school?

3

Marks

9. **(continued)**

The school ran a health promotion campaign to improve the fitness of its pupils.

The pie chart below shows how the pupils travelled to school after the campaign had been running for six months.

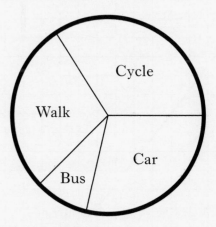

(*b*) Describe the differences in how the pupils travelled to school before and after the campaign.

2

[Turn over

Marks

10. The scale drawing shows the position of two islands in an ocean.

The scale drawing is **1 centimetre represents 100 kilometres**.

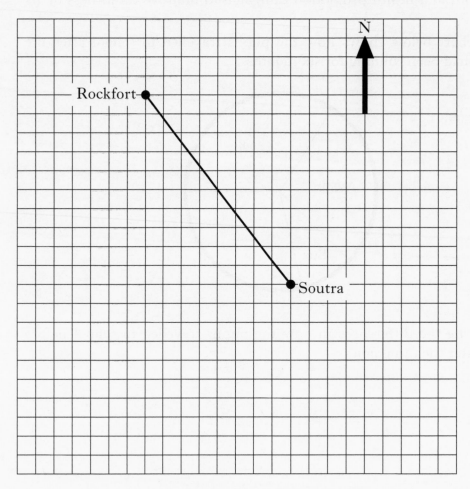

(*a*) Use the scale drawing to find the distance in kilometres from Rockfort to Soutra.

1

(*b*) Tern Island lies on a bearing of   • 180° from Rockfort
                                          • 233° from Soutra.

Complete the scale drawing above to show the position of Tern Island.

3

Marks

11. Last year Asim's income was £17 500.

This year his income increased to £18 200.

Calculate the increase as a percentage of last year's income.

4

[Turn over

Marks

12. The diagram below shows the position of a ceiling fan in a conservatory.

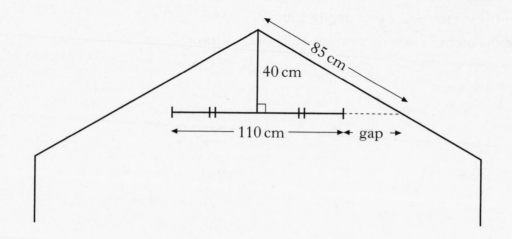

- The fan is 40 centimetres below the top of the conservatory.
- The sloping roof measures 85 centimetres to the level of the fan.
- The fan measures 110 centimetres across.

Calculate the size of the gap between the edge of the fan and the sloping roof.

Do not use a scale drawing.

4

Marks

13. The diagram below shows a semi-circular wall with a door.

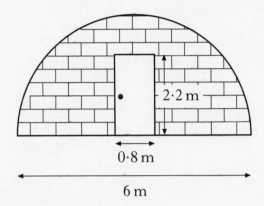

2·2 m

0·8 m

6 m

The door is a rectangle and is made of wood.

The rest of the wall is made of brick.

Calculate the area of brick wall in square metres.

Give your answer correct to one decimal place.

5

[END OF QUESTION PAPER]

ADDITIONAL SPACE FOR ANSWERS

2010

[BLANK PAGE]

FOR OFFICIAL USE

Total Mark

X101/102

NATIONAL QUALIFICATIONS 2010

FRIDAY, 21 MAY 1.00 PM – 1.35 PM

MATHEMATICS INTERMEDIATE 1
Units 1, 2 and
Applications of Mathematics
Paper 1
(Non-calculator)

Fill in these boxes and read what is printed below.

Full name of centre

Town

Forename(s)

Surname

Date of birth

Day	Month	Year

Scottish candidate number

Number of seat

1 **You may NOT use a calculator.**

2 Write your working and answers in the spaces provided. Additional space is provided at the end of this question-answer book for use if required. If you use this space, write clearly the number of the question involved.

3 Full credit will be given only where the solution contains appropriate working.

4 Before leaving the examination room you must give this book to the Invigilator. If you do not you may lose all the marks for this paper.

Use blue or black ink. Pencil may be used for graphs and diagrams only.

FORMULAE LIST

Circumference of a circle: $C = \pi d$
Area of a circle: $A = \pi r^2$
Curved surface area of a cylinder: $A = 2\pi rh$

Theorem of Pythagoras:

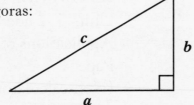

$$a^2 + b^2 = c^2$$

Marks

ALL questions should be attempted.

1. (*a*) Find $9 \cdot 22 - 5 \cdot 3$.

1

(*b*) Find $528 \div 300$.

1

(*c*) Find 60% of 250.

1

[Turn over

Marks

2. The graph shows the amount Megan spent each month on fruit and on sweets during 2009.

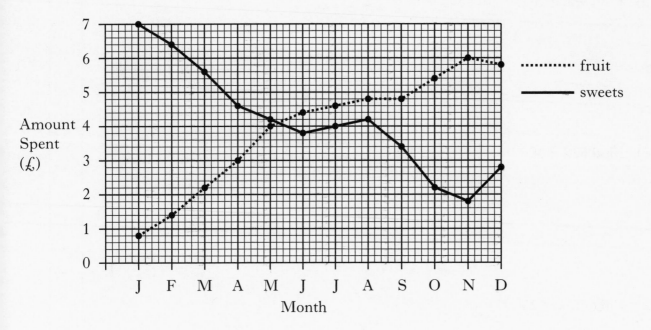

(a) How much did Megan spend on fruit in February?

1

(b) Describe the trend in the amount Megan spent on **both** fruit and sweets.

1

Marks

3. The network diagram shows the time it took for a racing car to make a pit stop. All times are in seconds.

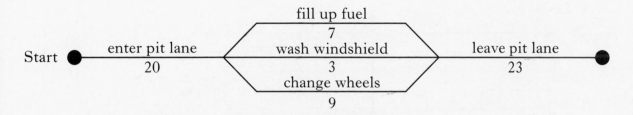

(*a*) How long did it take to change the wheels?

1

(*b*) How long did it take altogether for the pit stop from start to finish?

1

[Turn over

Marks

4. (*a*) On the grid below, plot the points A(−5,−2) and B(3,−2).

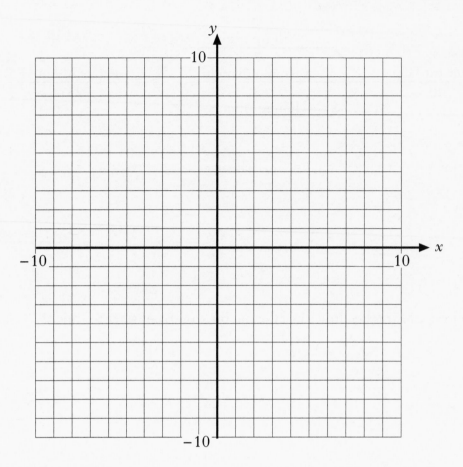

1

(*b*) Plot the point C so that triangle ABC is isosceles and has an area of 24 square units.

2

Marks

5. Malika wants to buy some home entertainment equipment from the items listed below.

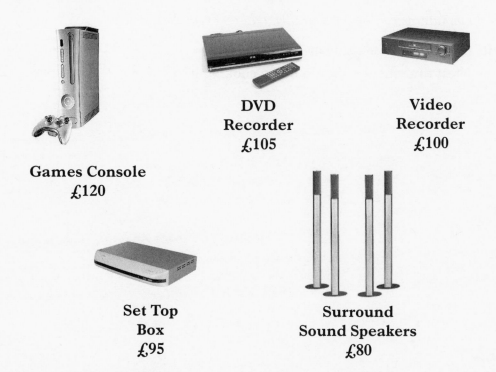

Games Console
£120

**DVD
Recorder
£105**

**Video
Recorder
£100**

**Set Top
Box
£95**

**Surround
Sound Speakers
£80**

Malika wants to buy three items.

She can afford to spend a maximum of £300.

She does not want to buy more than one of each item.

One combination of three items that Malika can buy is shown in the table below.

Games Console £120	DVD Recorder £105	Video Recorder £100	Set Top Box £95	Surround Sound Speakers £80	Total Value
	✓		✓	✓	£ 280

Complete the table to show **all** possible combinations of three items that Malika can buy.

3

Marks

6. Tom is going to cook a 3·5 kilogram turkey.

   He uses this rule to calculate the cooking time:

   "Cook for 40 minutes per kilogram and then add an extra 25 minutes."

   Tom wants the turkey to be ready at 1.30 pm.

   What is the latest time that he should begin cooking it?

   4

Marks

7. The number of hours of sunshine in Lerwick each December during a 10 year period is listed below.

16   20   13   11   28   16   31   24   15   23

Complete the boxplot, drawn below, to show the number of hours of sunshine.

11   24

4

[Turn over

Marks

8.   Zoe is finding the height of a building.

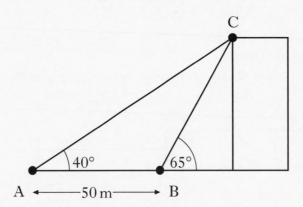

Points A and B are 50 metres apart.

Point C is at the top of the building.

The angle of elevation of Point C is • 40° from Point A.

  • 65° from Point B.

The position of Point A is shown on the grid below.

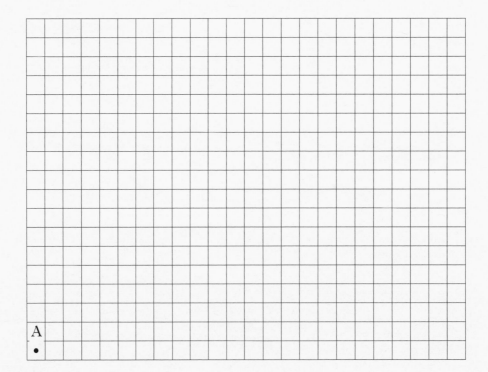

(a) Make a scale drawing to show the positions of Points B and C.

Use a scale of **1 cm to 10 m**.

2

Marks

8. (continued)

(*b*) Use your scale drawing to find the actual height of the building.

2

[Turn over for Question 9 on *Page twelve*

Marks

9. The rules to complete a number pyramid are:
   - the number in a circle is equal to the two numbers in the circles immediately below it multiplied together.
   - only positive and negative whole numbers can be used.

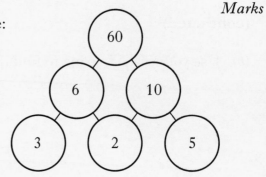

(*a*) Use the rules to complete this number pyramid.

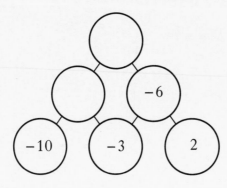

2

(*b*) Use the rules to complete this number pyramid.

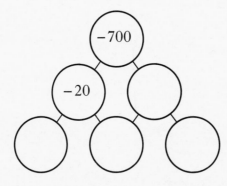

3

[END OF QUESTION PAPER]

ADDITIONAL SPACE FOR ANSWERS

DO NOT
WRITE I
THIS
MARGIN

ADDITIONAL SPACE FOR ANSWERS

FOR OFFICIAL USE

Total Mark

X101/104

NATIONAL QUALIFICATIONS 2010

FRIDAY, 21 MAY 1.55 PM – 2.50 PM

MATHEMATICS
INTERMEDIATE 1
Units 1, 2 and
Applications of Mathematics
Paper 2

Fill in these boxes and read what is printed below.

Full name of centre

Town

Forename(s)

Surname

Date of birth

Day   Month   Year

Scottish candidate number

Number of seat

1 **You may use a calculator.**

2 Write your working and answers in the spaces provided. Additional space is provided at the end of this question-answer book for use if required. If you use this space, write clearly the number of the question involved.

3 Full credit will be given only where the solution contains appropriate working.

4 Before leaving the examination room you must give this book to the Invigilator. If you do not you may lose all the marks for this paper.

Use blue or black ink. Pencil may be used for graphs and diagrams only.

FORMULAE LIST

Circumference of a circle: $C = \pi d$

Area of a circle: $A = \pi r^2$

Curved surface area of a cylinder: $A = 2\pi r h$

Theorem of Pythagoras:

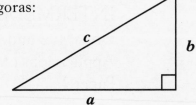

$a^2 + b^2 = c^2$

DO NOT
WRITE IN
THIS
MARGIN

Marks

ALL questions should be attempted.

1. A car travelling at an average speed of 80 kilometres per hour takes 2 hours 45 minutes for the journey from Dundee to Inverness.

    Calculate the distance between the two towns.

    2

2. Tanya takes out a life insurance policy worth £45 000.

    The insurance company charges a monthly premium of £1·30 for every £1000 worth of cover.

    How much will Tanya pay **annually** for this policy?

    2

[Turn over

Marks

3. The table below shows the **monthly payments** to be made when money is borrowed from a bank.

Borrowers can choose to make payments with or without payment protection.

	Amount Borrowed					
	£30 000		£40 000		£50 000	
Period of loan	without payment protection	with payment protection	without payment protection	with payment protection	without payment protection	with payment protection
25 years	£243	£292	£325	£389	£406	£487
20 years	£262	£314	£349	£419	£437	£524
15 years	£297	£356	£396	£475	£495	£593
10 years	£373	£447	£498	£597	£622	£746

(a) Lucy borrows £40 000 over 15 years **without payment protection**.
State her monthly payment.

1

(b) Over the 15 years, how much **extra** would Lucy pay **in total** for payment protection on her loan of £40 000?

2

Marks

4. A group of people are booking flights to Munich.

This flowchart is used to calculate the total cost of the flights.

There are **8** people in the group. They want to fly on a **Friday**.

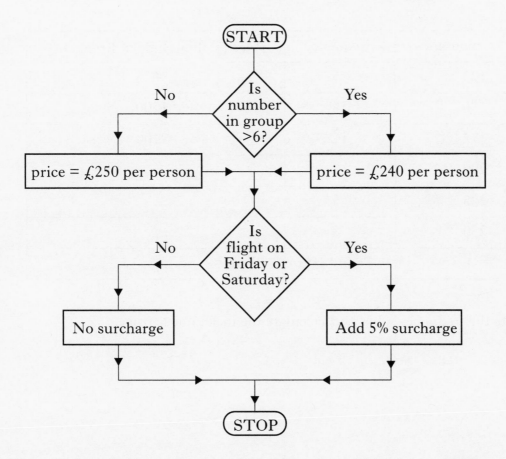

Calculate the **total** cost for these **8** people to fly to Munich on a **Friday**.

2

[Turn over

Marks

5. An estate agency recorded the prices of the houses they sold in April.

   The prices varied from £125 000 to £250 000.

   The prices are shown in the frequency table below.

Price (£ thousands)	Frequency	Price (£ thousands) × Frequency
125	5	625
150	8	1200
175	12	2100
200	7	
225	5	
250	3	
	Total = 40	Total =

Complete the frequency table **and** calculate the mean house price.

3

Marks

6. Jamie is a cleaner.

   He works Monday to Friday from 6 am until 8 am and from 3.30 pm until 6 pm.

   His basic rate of pay is £6·80 per hour.

   (*a*) Calculate his weekly wage.

2

   (*b*) Jamie was paid at time and a half for working one Saturday.
   His wage for the day was £51.
   How many hours did he work that day?

2

[Turn over

Marks

7. The screen size of a laptop computer is the length of the diagonal from one corner of the rectangular screen to its opposite corner.

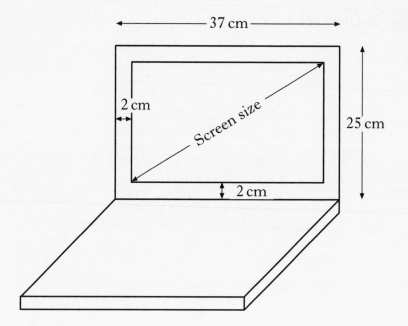

This laptop measures 37 centimetres by 25 centimetres as shown.

The frame around the screen has a width of 2 centimetres.

Calculate the screen size of this laptop.

Do not use a scale drawing.

4

Marks

8. David bought a computer game in the United States for 50 dollars.

   The same game cost £35 in Scotland.

   The exchange rate was £1 = $1·62.

   How much did David save by buying the game in the United States?

   Give your answer in pounds and pence.

   3

9. Charlie invests £4200 in a bank account.

   The rate of interest is 1·3% per annum.

   Calculate the interest he should receive after 9 months.

   3

   [Turn over

Marks

10. This cuboid has a square base.

   Its height is 25 centimetres and its volume is 1369 cubic centimetres.

   Calculate the length of its base.

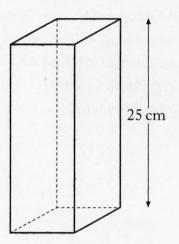

25 cm

3

11. Tony sells jewellery.

   One day he earned £90 commission for selling jewellery worth £750.

   Express Tony's commission as a percentage of his sales.

3

Marks

12. Two classes of fourteen pupils at Oakland Academy collected money for a local charity.

Listed below are the amounts (in £) collected by the pupils in class 5C.

27 26 17 27 18 21 23 19 18 27 24 20 31 28

(a) Find the median.

2

(b) Find the range.

1

(c) For class 5M the median was £10 and the range was £17.

Make **two** comments comparing the amounts collected by the pupils in class 5C and class 5M.

2

[Turn over

Marks

13. A plant container is in the shape of a cylinder with diameter 12 centimetres and height 20 centimetres.

The container is closed at the bottom and open at the top.

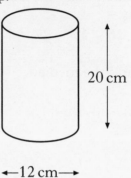

20 cm

←12 cm→

The diagram below shows the net of the container.

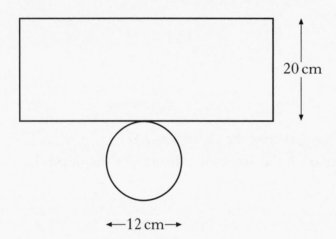

20 cm

←12 cm→

Calculate the **total** surface area of the container.

4

Marks

14. A sign for a mushroom farm consists of a semi-circle and a rectangle.

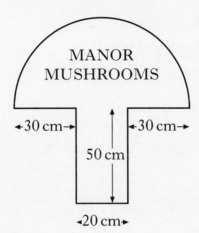

There is a red border painted all around the edge of the sign.

Calculate the total length of the red border.

Give your answer correct to the **nearest centimetre**.

5

[Turn over for Question 15 on *Page fourteen*

Marks

15. A box contains 3 red pencils and 12 green pencils.

(*a*) A pencil is taken from the box.
What is the probability that the pencil is red?
Give your answer as a fraction in its simplest form.

2

(*b*) The pencil is put back in the box.
More red pencils are then added to the box.
The probability of taking a red pencil is now $\frac{1}{3}$.
How many red pencils are now in the box?

2

[END OF QUESTION PAPER]

DO NOT
WRITE IN
THIS
MARGIN

ADDITIONAL SPACE FOR ANSWERS

ADDITIONAL SPACE FOR ANSWERS